SCIENTIFIC AMERICAN EDUCATIONAL PUBLISHING

SCIENTIFIC AMERICAN INVESTIGATES THE HUMAN BODY

THE CIRCULATORY SYSTEM

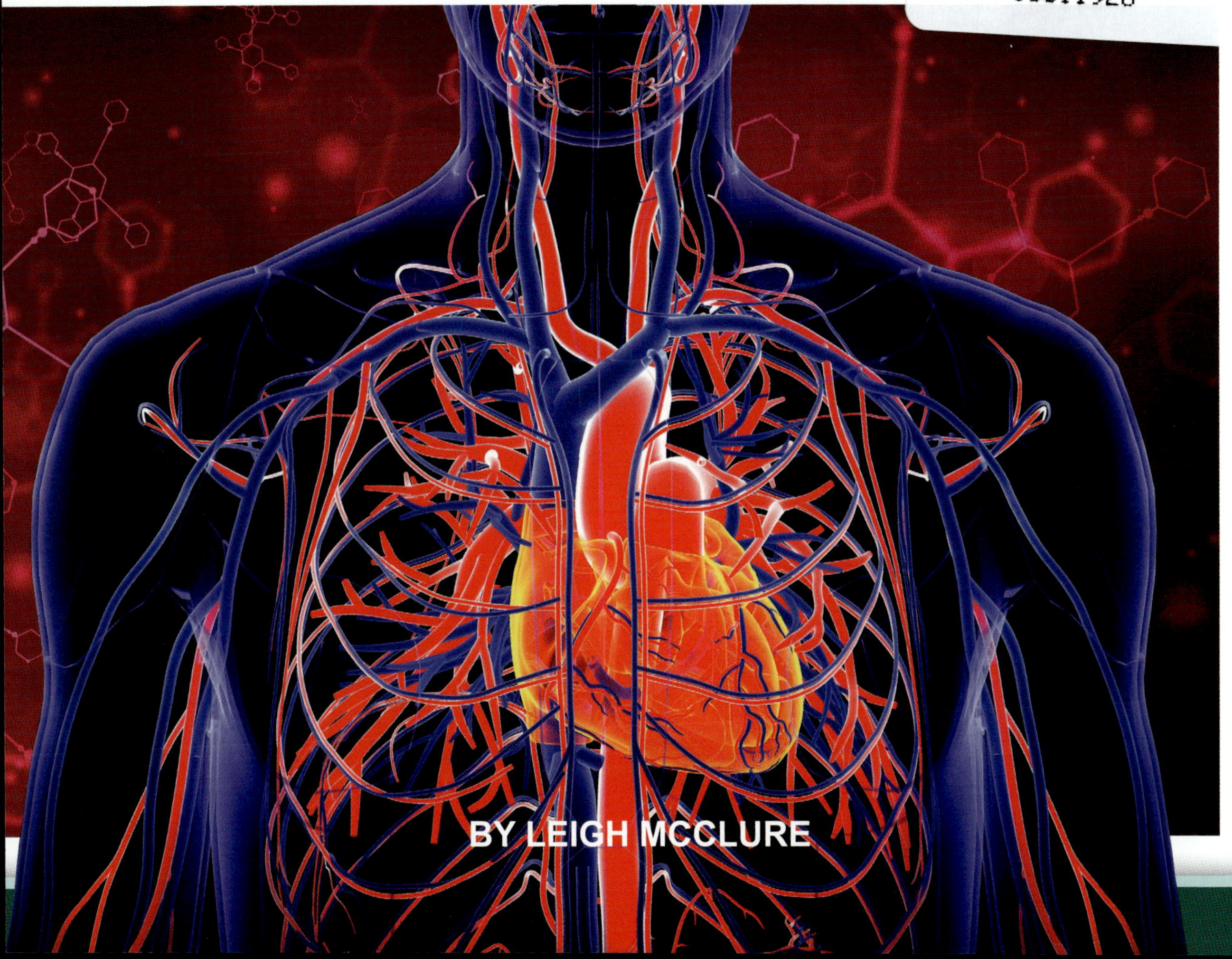

BY LEIGH MCCLURE

Published in 2025 by The Rosen Publishing Group
in association with Scientific American Educational Publishing
2544 Clinton Street, Buffalo NY 14224

Cataloging-in-Publication Data
Names: McClure, Leigh.
Title: The circulatory system / Leigh McClure.
Description: New York : Scientific American Educational Publishing, an imprint of Rosen Publishing, 2025. | Series: Scientific American Investigates the Human Body | Includes glossary and index.
Identifiers: ISBN 9781725351356 (pbk.) | ISBN 9781725351363 (library bound) | ISBN 9781725351370 (ebook)
Subjects: LCSH: Cardiovascular system–Juvenile literature.
Classification: LCC QM178.M34 2025 | DDC 612.1–dc23

Designer: Andrea Davison-Bartolotta
Editor: Caitie McAneney

Portions of this work were originally authored by Autumn Leigh and published as *The Circulatory System*. All new material in this edition authored by Leigh McClure.

Photo credits: Cover, p. 1 (main) Shot4Sell/Shutterstock.com; cover, p. 1 (background) jijomathaidesigners/Shutterstock.com; p. 5 (inset) Net Vector/Shutterstock.com; p. 5 (main) aDam Wildlife/Shutterstock.com; pp. 7, 17 Komsan Loonprom/Shutterstock.com; p. 9 Katalin Macevics/Shutterstock.com; pp. 11, 27 PeopleImages.com - Yuri A/Shutterstock.com; p. 13 DiBtv/Shutterstock.com; p. 15 sciencepics/Shutterstock.com; p. 19 Rybalchenko Nadezhda/Shutterstock.com; p. 20 cenksns/Shutterstock.com; p. 21 Eva Foreman/Shutterstock.com; p. 23 Antonio Guillem/Shutterstock.com; p. 25 Pixel-Shot/Shutterstock.com; p. 29 ntkris/Shutterstock.com.

Printed in the United States of America

CPSIA compliance information: Batch #CSSA25. For Further Information contact Rosen Publishing at 1-800-237-9932.

CONTENTS

Words in the glossary appear in **bold** type the first time they are used in the text.

BLOOD ON THE MOVE

Have you ever wondered how blood gets around your body? The movement of blood is thanks to the circulatory system. This body system is made up of the blood, blood vessels, and heart. These parts work together to help the body work the way it's supposed to.

Circulation means the movement of blood through the body. However, blood carries more than you might think. It transports chemicals, **nutrients**, oxygen, and water throughout the body. Organs and tissues need these things to work. It also moves some waste products out of the body.

Humans have a closed circulatory system. That means blood is transported through blood vessels to reach organs. This is different from an open circulatory system, in which blood surrounds the cells directly.

Humans and other **mammals** have a closed circulatory system. **Arthropods** have an open circulatory system.

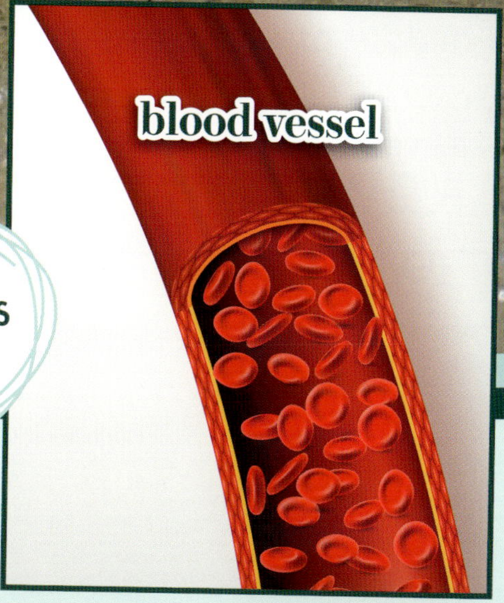

FUN FACT

"CIRCULATE" MEANS TO MOVE AROUND WITHOUT STOPPING. THAT'S WHERE THE CIRCULATORY SYSTEM GETS ITS NAME.

THE STAR OF THE SYSTEM

When it comes to the circulatory system, the heart is the star of the show. This organ is made up of muscle and tissue. The heart makes sure blood is always moving through the blood vessels, even when you're asleep. This allows the circulatory system to continuously supply your cells and organs with fresh nutrients and oxygen. It also removes harmful waste products, like carbon dioxide.

Your heart is about the size of your fist. It's located in the middle of your chest, between your lungs and slightly to the left side of your body. The rib cage keeps it safe. When you put your hand up to your chest, you can feel your heart beating. What you're feeling is your heart squeezing and relaxing very quickly.

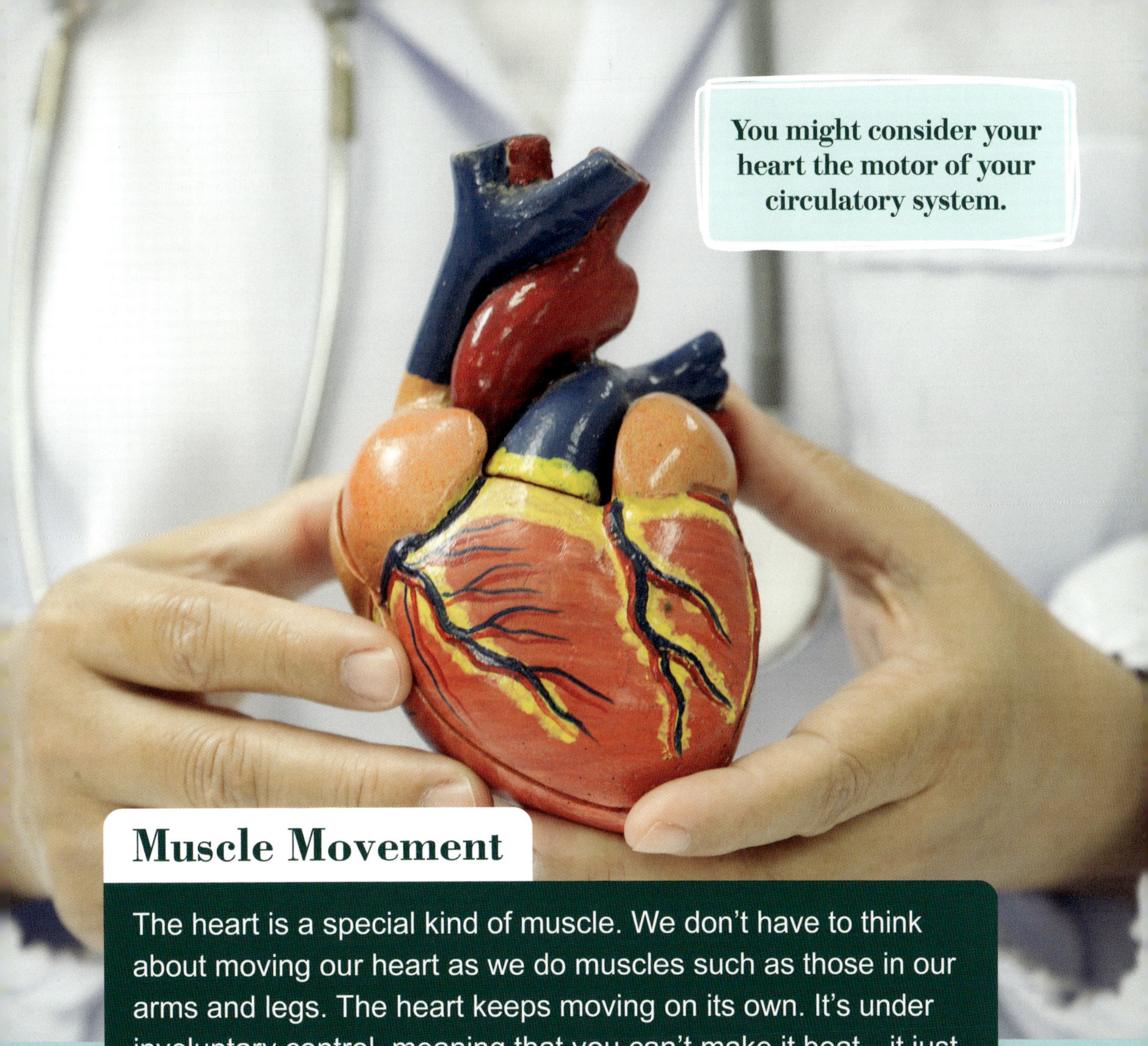

You might consider your heart the motor of your circulatory system.

Muscle Movement

The heart is a special kind of muscle. We don't have to think about moving our heart as we do muscles such as those in our arms and legs. The heart keeps moving on its own. It's under involuntary control, meaning that you can't make it beat—it just happens. This movement is what keeps your blood flowing. A single heartbeat lasts about 0.8 second. Every day, about 2,000 gallons (7,571 L) of blood pass through the human heart.

PUMP IT UP

Hearts are **complex** organs. Each human heart has four chambers arranged on two sides. The upper and lower chambers on each side form a pump. Each upper chamber is called an atrium. Each lower, larger chamber is called a ventricle. The left ventricle is the largest and strongest chamber of the heart. It needs to be strong to pump blood throughout the body.

The atrium on the right side of the heart receives blood from the body. Blood passes from the atrium into the right ventricle and is pumped to the lungs. The left atrium receives blood from the lungs, and the left ventricle pumps it to the rest of the body. Your heartbeat is produced by the atria contracting, then ventricles contracting.

As you read about the different parts of the heart, point to them on this diagram. Trace the flow of blood.

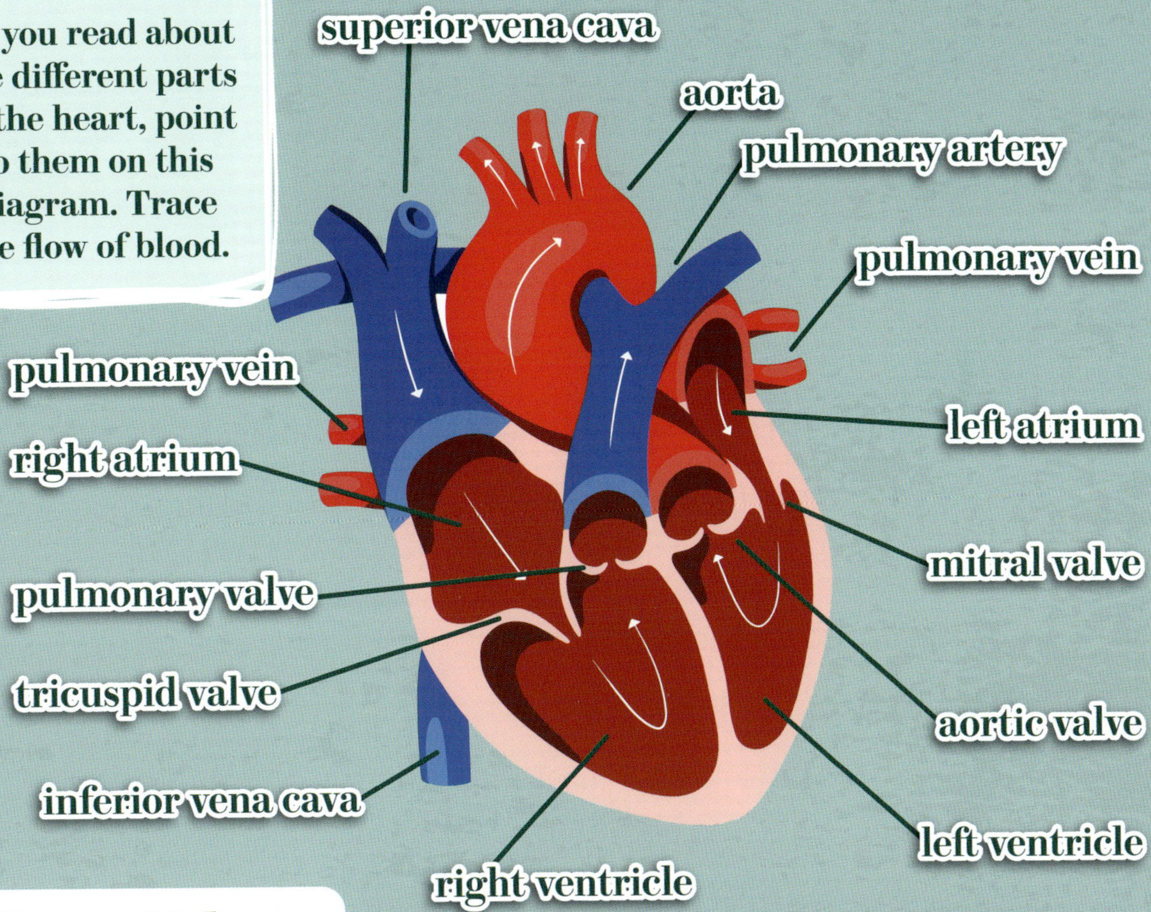

Four Valves

Blood flow is controlled by four **valves** in the heart. The aortic valve allows blood to pass from the left ventricle out to the rest of your body. The tricuspid valve controls blood flow between the right atrium and right ventricle. The **pulmonary** valve allows blood to flow from the right ventricle out to the lungs to get oxygen. The mitral valve sends oxygen-rich blood from the lungs through the left atrium and into the left ventricle.

WHAT ARE BLOOD VESSELS?

Blood vessels are like a highway for blood. These hollow tubes carry blood where it needs to go. The three kinds of blood vessels are arteries, veins, and capillaries. As the vessels extend farther from the heart, they branch out into smaller tubes.

Arteries carry blood away from the heart. The aorta, which begins at the left ventricle, is the largest artery. It rises out of the top of the heart, turns downward, and extends just below the **diaphragm**. Femoral arteries are large blood vessels that run through the thighs. Carotid arteries travel up the neck and provide the brain with oxygenated blood. The pulmonary arteries carry blood from the heart to the lungs to get oxygen. They're the only arteries that carry deoxygenated blood, or blood from which oxygen has been removed.

FUN FACT

ADULT HUMANS HAVE ABOUT 60,000 MILES (96,560 KM) OF BLOOD VESSELS IN THEIR BODY! LAID END TO END, THEY WOULD REACH AROUND EARTH MORE THAN TWICE.

Many people check their heart rate when exercising. The more you work your body, the faster your heart will beat.

Feel the Beat

Take a moment to feel your heart beating. Your heartbeat can be felt as a pulse by pressing on certain arteries. Two of these are the carotid arteries on either side of your neck. Count how many times your heart beats in one minute. That will give you your heart rate. You can also feel your pulse in the radial arteries of your wrists. Try doing jumping jacks and then taking your pulse again. What do you notice?

While arteries mostly carry oxygen-rich blood, veins are highways for oxygen-poor blood. This blood needs to go from the tissues of the body back to the heart. Veins are thinner than arteries and not as strong. Since the veins carry the body's waste products, it's important that blood isn't allowed to flow backward in them. To stop this from happening, veins have tiny valves to keep blood flowing toward the heart. The largest veins are the superior vena cava and inferior vena cava. These veins connect directly to the heart.

Capillaries connect the arteries to the veins. They're the smallest of the blood vessels. They are so thin that carbon dioxide passes from body tissue into the capillaries. Oxygen and nutrients also pass through capillary walls and into the surrounding tissues.

The arterial system is made up of arteries, while the venous system is made up of veins.

Circulatory System

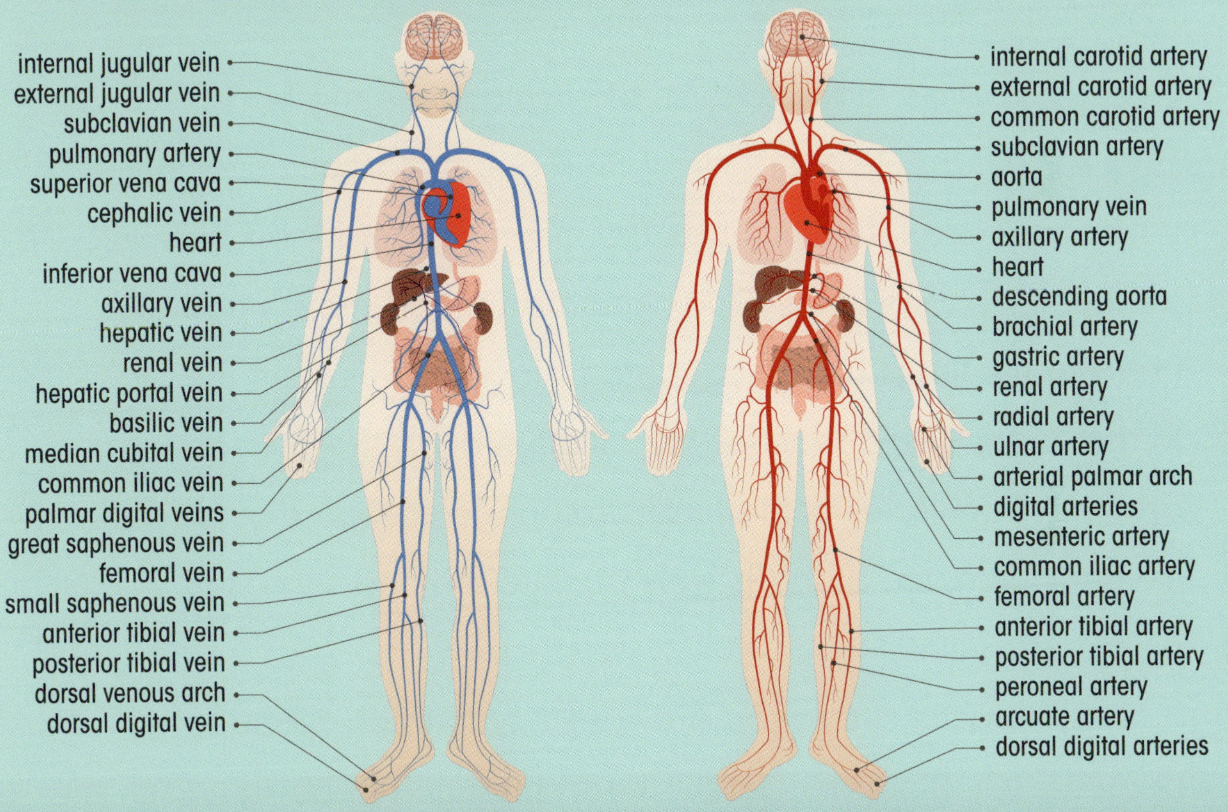

FUN FACT

THE LARGEST BLOOD VESSEL, THE AORTA, CAN BE MORE THAN 1 FOOT (30.5 CM) LONG.

CIRCULATION STATION

Let's follow the two types of circulation in the body. They're called pulmonary circulation and systematic circulation. Pulmonary circulation involves the movement of blood between the heart and the lungs. Deoxygenated blood moves from the heart to the lungs and oxygenated blood moves from the lungs to the heart. Blood reaches the lungs through the pulmonary arteries. Capillaries in the lungs exchange carbon dioxide for oxygen. Then, the blood travels back to the heart through the pulmonary veins.

Systematic circulation provides oxygenated blood to all body tissues. Blood leaves the heart through the aorta, which branches out into increasingly smaller arteries. It passes through tiny capillaries, then releases oxygen. The deoxygenated blood travels through the veins back to the heart.

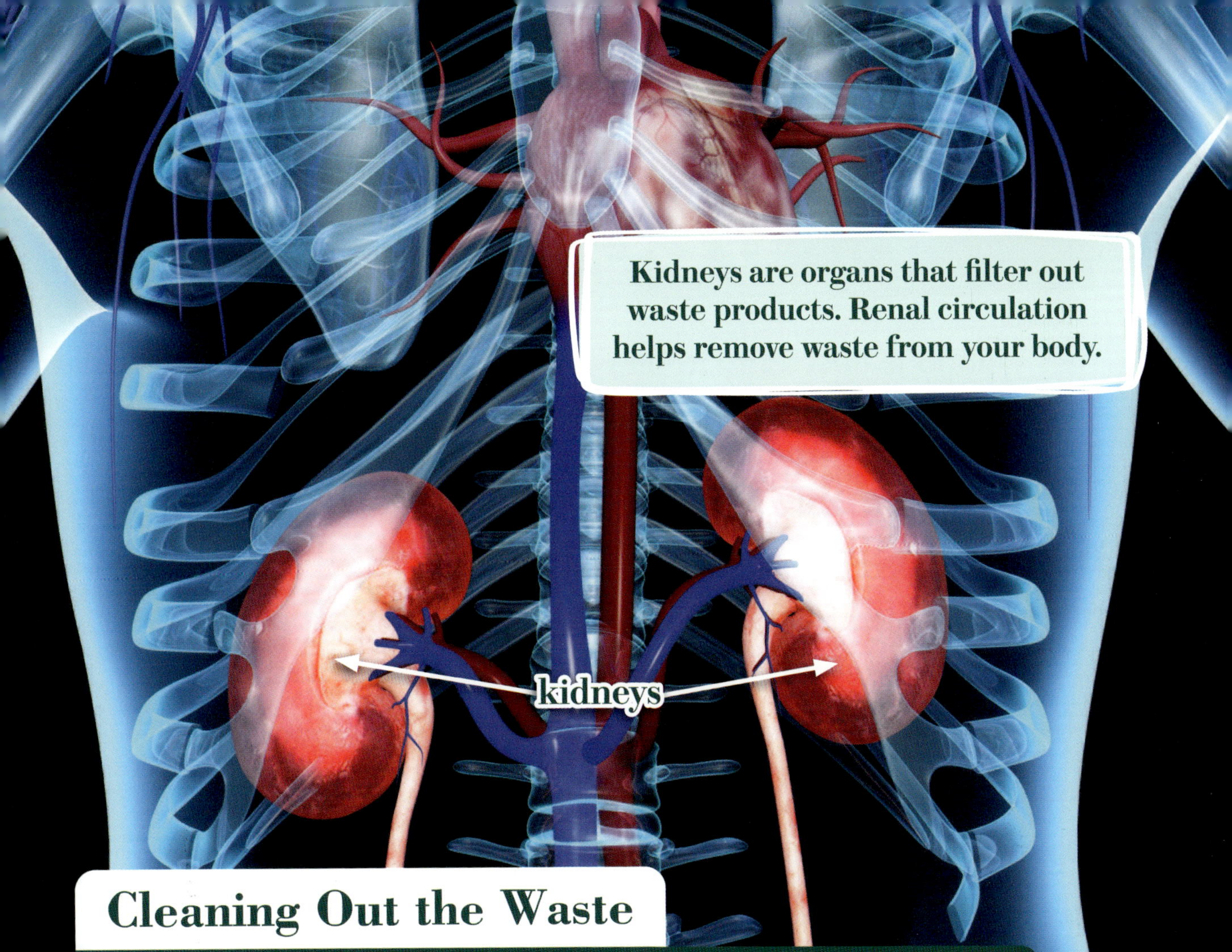

Kidneys are organs that filter out waste products. Renal circulation helps remove waste from your body.

Cleaning Out the Waste

One important kind of systematic circulation is renal circulation, which has to do with the kidneys. Oxygen-rich blood enters the kidneys through the renal arteries. These arteries branch out into capillaries, which take the blood to the parts of the kidneys that filter out waste products. The kidneys make that waste into **urine** to leave your body. Clean blood can then go back to the heart, thanks to systematic circulation.

ALL ABOUT BLOOD

Blood is necessary to all workings of the human body. It brings oxygen and nutrients to every cell. It also carries waste products away. The average adult has approximately 5.3 quarts (5 L) of blood. Blood makes up 7 to 8 percent of a person's body weight!

Blood is made up of two parts: cells and a clear, yellowish liquid called plasma. The cells in our blood include red blood cells, white blood cells, and platelets. Each cell has a different function, or job. A little more than half of blood is plasma. About 90 percent of plasma is water. The other 10 percent is made up of waste products and nutrients. This includes proteins—nutrients our bodies need to build, maintain, and replace tissues.

FUN FACT

LYMPH INCLUDES HORMONES, WHICH ARE CHEMICALS THAT AFFECT THE WAY OUR BODIES FUNCTION.

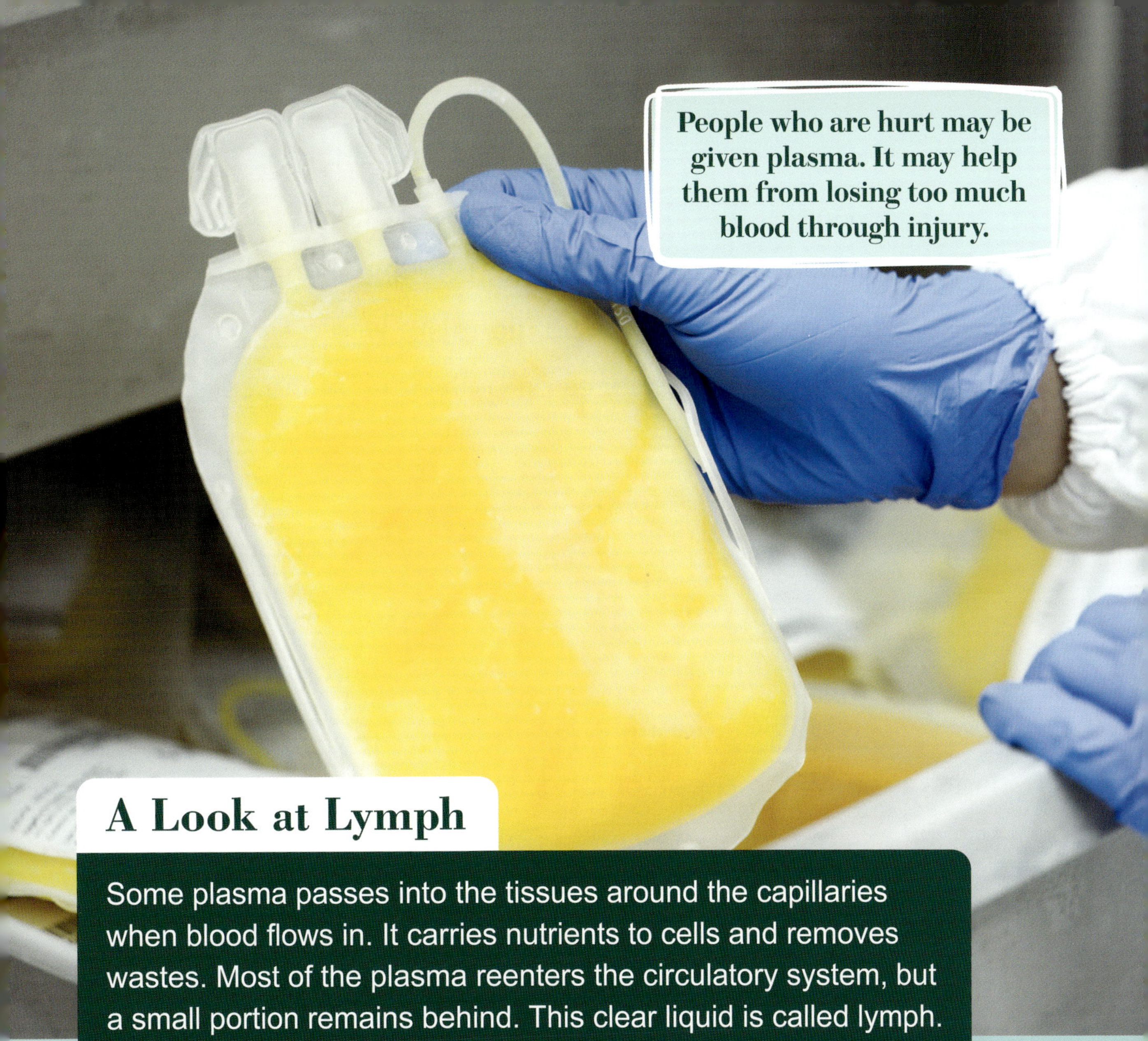

People who are hurt may be given plasma. It may help them from losing too much blood through injury.

A Look at Lymph

Some plasma passes into the tissues around the capillaries when blood flows in. It carries nutrients to cells and removes wastes. Most of the plasma reenters the circulatory system, but a small portion remains behind. This clear liquid is called lymph. Lymph travels toward the upper chest through the vessels that make up the lymphatic system. Wastes in lymph are filtered out by small lumps of tissue called nodes. The filtered lymph goes back into the circulatory system.

Each kind of blood cell has an important job to do. Red blood cells (RBCs) carry oxygen from the lungs to the rest of the body. RBCs are round, flat disks with a biconcave shape. This means the surfaces on both sides of the cell curve inward.

As blood passes through the lungs, oxygen attaches itself to a molecule in RBCs called hemoglobin. Oxygen-rich blood is bright red, and oxygen-poor blood is dark red. RBCs wear out after about 120 days. They're replaced by new cells. Old RBCs are removed from the body by the **liver** and **spleen**. When a person doesn't have enough healthy RBCs, they can develop anemia. This illness makes people feel tired and weak.

FUN FACT

IN CHILDREN, BLOOD CELLS ARE MADE IN THE BONE MARROW OF MANY BONES. IN ADULTS, BLOOD CELLS ARE PRODUCED IN THE MARROW OF THE SPINE, BREASTBONE, RIBS, PELVIS, UPPER ARMS, UPPER LEGS, AND SKULL.

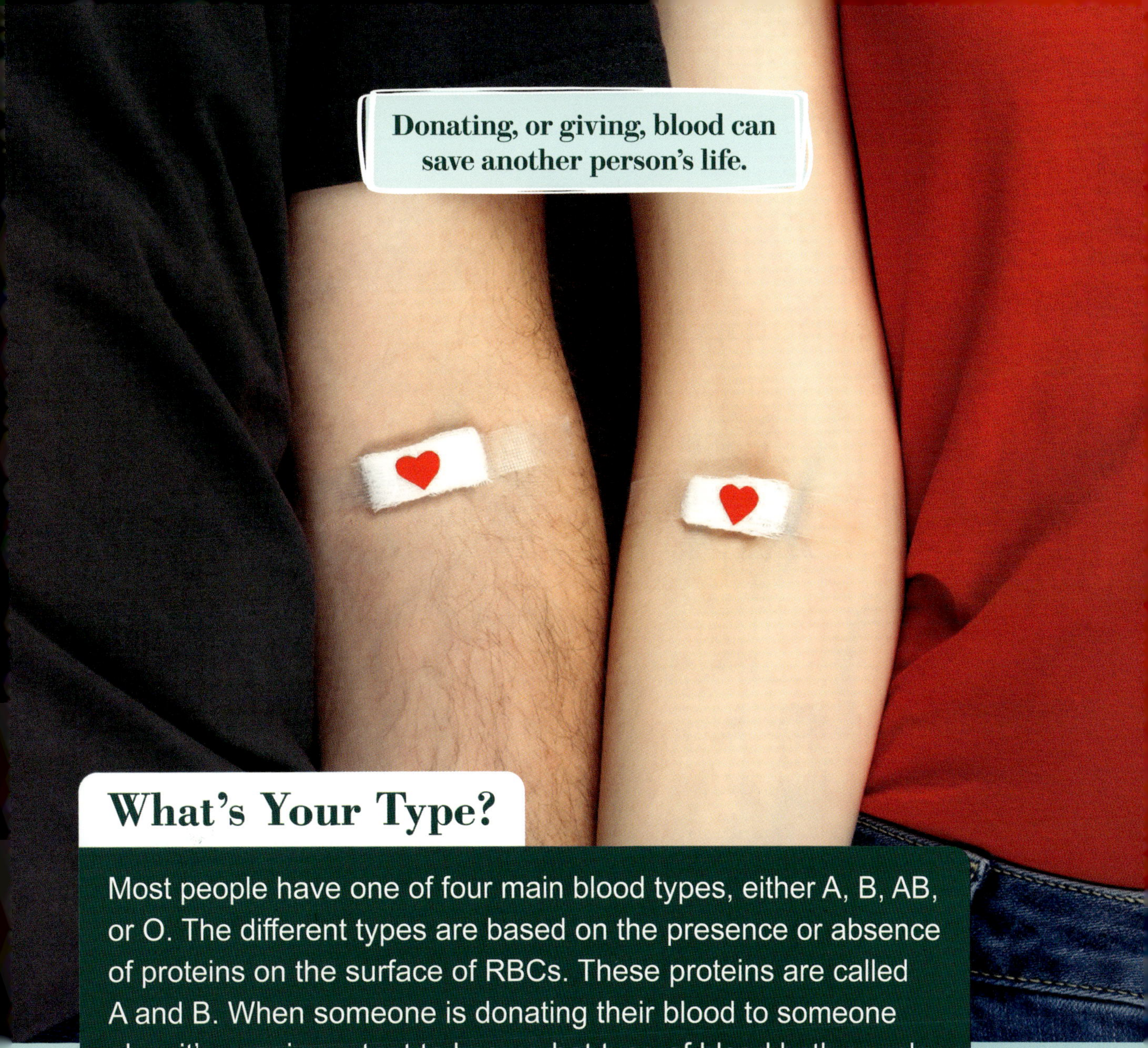

Donating, or giving, blood can save another person's life.

What's Your Type?

Most people have one of four main blood types, either A, B, AB, or O. The different types are based on the presence or absence of proteins on the surface of RBCs. These proteins are called A and B. When someone is donating their blood to someone else, it's very important to know what type of blood both people have or problems can occur. People with type O blood can donate blood to anyone. People with AB blood can only donate to others with the same blood type.

While red blood cells deliver oxygen, white blood cells (WBCs) fight illness. Everyone gets infections from time to time. An infection happens when germs enter the body and then increase in number. Germs are tiny living things that can only be seen under a microscope. Many common types of germs are bacteria. Some kinds of bacteria make people sick. WBCs help fight "bad" bacteria and other germs that enter the body.

There are five major types of WBCs. Each has a different job. Some only go after a specific kind of germ. They surround the germs and "eat" them. Other WBCs use proteins called antibodies to attack germs. Depending on the kind, WBCs live from a few days to a few weeks. WBC production increases when germs appear.

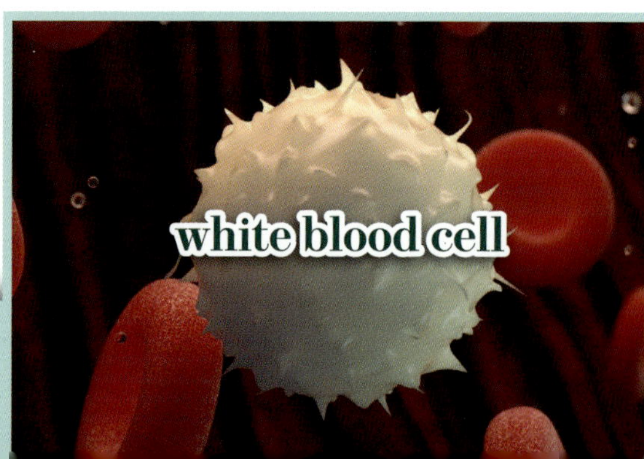

In people with **allergies**, WBCs attack normally harmless substances, such as dust, pollen, and mold.

Platelets to the Rescue!

Have you ever had a scab after getting a cut? You can thank your platelets! When blood vessels are damaged or cut, oval cells called platelets rush to the injury. The platelets branch out and fit together like puzzle pieces to seal the wound. These tiny cells use a protein to help them stick together. The clump that forms is called a clot. It makes a plug that keeps blood inside the vessels and germs outside.

HORMONE TALK

Hormones are another important part of the human body. These "chemical messengers" give organs information about how to function. This happens all the time in the human body. Hormones wouldn't be able to get where they need to go without the circulatory system.

In the human body, organs called glands make hormones and secrete, or release, them into the bloodstream. There are numerous glands in the human body. Some of the major glands include the thyroid, pancreas, adrenal glands, and pituitary glands. Together, a number of organs and glands in the human body make up the endocrine system. The word "endocrine" comes from the Greek words for "inside" and "secrete." Under certain conditions, the heart also secretes a hormone that helps regulate blood flow and kidney function.

The "fight-or-flight response" is the human body's inborn ability to prepare for a fight or run away.

Under Threat

What happens to your body when you see a threat? The brain first sends messages to several parts of the body, including glands. Adrenaline is a hormone from the adrenal glands that tells the heart to beat faster. It causes the arteries to grow narrower and the veins to grow wider. This increases blood flow throughout the body. The pituitary gland releases endorphins—natural painkillers—into the bloodstream. This gets you ready to either fight the threat or flee from it.

HEART DISEASE

Because the circulatory system is so important, any disorder can cause big problems in the body. Some affect the heart, while others affect the blood vessels and blood. "Heart disease" is a general term for any disease that affects the blood vessels and heart. Blood pressure is a measurement of the force of blood pumping through the arteries. Hypertension (high blood pressure) and hypotension (low blood pressure) are signs of heart disease.

The most common form of heart disease is coronary artery disease. Fatty material builds up on artery walls, causing them to become narrow and hardened. Blood flow to the heart is reduced. This can cause chest pain and heart attack. Heart disease also includes an uneven heartbeat and heart failure. Eating too much of certain kinds of unhealthy foods and smoking can lead to heart disease.

FUN FACT

SINCE 1950, HEART DISEASE HAS BEEN THE LEADING CAUSE OF DEATH IN ADULTS IN THE UNITED STATES.

A heart attack can result in lifelong heart damage or death. It's important that someone gets help right away.

Stents Save Lives

A simple medical device called a stent can save the lives of patients with coronary artery disease. A stent is a small, wire-mesh tube designed to widen blocked arteries and restore normal blood flow. A balloon is placed inside the stent. Doctors make a small cut in an arm or leg artery and then slide the stent along the artery to the blockage. The balloon is filled with air, which causes the stent—as well as the artery—to open up.

BLOOD DISORDERS

Some people have genetic blood disorders, or those that are passed down from parent to child. Other people develop blood disorders over time. Abnormal hemoglobin may cause sickle cell anemia, a genetic disorder. People with sickle cell anemia deal with pain, low energy, and weakness.

Some blood disorders make people more likely to have blood clots. A pulmonary embolism is a sudden, life-threatening blockage in a lung artery. It's usually caused by a blood clot that travels from a vein in a leg to the lungs.

People with hemophilia, a rare disorder, have blood that doesn't clot properly. They bleed longer than other people and can suffer from internal bleeding. Most cases of hemophilia are passed from parent to child. People with hemophilia need to be careful to avoid injury.

Doctors can look at the shape and amount of blood cells to tell if a person has a blood disorder.

Cancer of the Blood

Leukemia is a **cancer** that affects the bone marrow and blood. Bone marrow in people with leukemia produces a large number of WBCs that don't act like normal cells. In time, these abnormal WBCs crowd out healthy cells in the bloodstream, lymph nodes, liver, and spleen. Leukemia causes anemia, bleeding, and infections. Some forms of leukemia can be cured, some can be controlled, and others are deadly.

HELPING YOUR HEART

Heart disease and blood disorders may seem scary. However, there are ways that you can help your heart. Have you ever heard someone say that exercise "gets the blood pumping"? Exercising regularly helps to improve blood flow and keep the circulatory system strong. It helps fight high blood pressure too.

Be mindful of what you put in your body. Smoking and taking drugs can cause circulation issues and heart disease. Foods with too much fat and sugar can lead to coronary artery disease. Hypertension is often made worse by salty foods, such as chips. However, eating a diet with lots of fruits, vegetables, protein, and grains can help your heart. You only get one heart, so it's good to treat it with the care it deserves!

Choose lean meats like chicken for your meals. That's a heart-healthy choice!

GLOSSARY

allergy: A sensitivity to normally harmless things in the environment, such as dust, pollen, or mold.

arthropod: An animal that lacks a backbone and has a skeleton on the outside of its body, such as an insect, spider, shrimp, or crab.

bone marrow: Soft matter inside bones where blood cells are made.

cancer: A disease caused by the uncontrolled growth of cells in the body.

complex: Made up of many different parts.

diaphragm: A thin, curved muscle below the lungs.

liver: An organ that stores and cleans blood.

mammal: A warm-blooded animal that has a backbone and hair, breathes air, and feeds milk to its young.

nutrient: Something a living thing needs to grow and stay alive.

pulmonary: Relating to or affecting the lungs.

spleen: An organ that destroys old blood cells and stores blood.

urine: Yellowish liquid secreted by the kidneys that contains waste products.

valve: Something that controls the movement of liquids or gases through tubes or vessels.

FOR MORE INFORMATION

Books

Green, Alicia. *Circulatory System*. New York, NY: Children's Press, 2024.

Hicks, Dwayne. *Your Amazing Heart*. Buffalo, NY: PowerKids Press, 2023.

Troup, Roxanne. *The Circulatory System*. Minneapolis, MN: Kids Core, 2023.

Websites

Explore the Body
kidshealth.org/en/kids/center/htbw-main-page.html
Check out more body systems, including the immune system and respiratory system.

Human Body
www.ducksters.com/science/biology/humanbody.php
Discover more about the body, including its main structures and senses.

The Truth About Your Heart
kids.nationalgeographic.com/science/article/the-truth-about-your-heart
Learn interesting facts about your heart with National Geographic Kids.

INDEX